Written by
Beth Davis

Cover Design by
Matthew Van Zomeren

Inside Art by
Sherry Neidigh

Publishers
Instructional Fair • TS Denison
Grand Rapids, Michigan 49544

About the Book

The Inquiry Science series was designed and tested by classroom teachers familiar with the National Science Education Standards. It is the goal of the series to apply the standards in a user-friendly format.

Promote minds-on learning by challenging the students to verbalize their observations and make inferences. Ask simple questions such as the following: What just happened? Why do you think that happened? What did you discover? Where have you seen that before? The science process skills used in each lesson are provided so you can raise students' awareness and highlight their importance.

Credits

Author: Beth Davis
Cover Design: Matthew Van Zomeren
Inside Illustrations: Sherry Neidigh
Project Director/Editor: Elizabeth Flikkema
Editors: Wendy Roh Jenks, Linda Triemstra
Page Design: Pat Geasler

About the Author

Beth Davis is an outstanding educator from Miami, Florida, where she is currently a science lab teacher for grades 2–5. Ms. Davis received her Bachelor of Science in Elementary Education from Florida International University. Her master's degree was earned at Nova University. Beth Davis has written curriculum and several articles in the areas of math and science. Her other Instructional Fair • TS Denison book is titled *Flowering Plants* and is also found in the Inquiry Science series.

Standard Book Number : 1-56822-677-2
Matter—Grades 2–3

a division of Tribune Education
2400 Turner Avenue NW
Grand Rapids, Michigan 49544

Table of Contents

What's the Matter?

Gearing Up

Get the students thinking by holding up an empty resealable plastic bag. Explain that there is something in this bag. Ask the students to tell you what they think is in the bag. Listen to their responses and tell them that in this exploration, they will discover what is in the bag. They will also examine the three states of matter—solid, liquid, and gas.

Process Skills Used

- observing
- predicting
- recording data
- comparing
- classifying

Guided Discovery

Background information for the teacher:
Matter is anything that has weight and takes up space. The three states of matter are solids, liquids, and gases. Write the following properties of the states of matter on the chalkboard. *Solids:* You can sec solids and they have a specific shape. You cannot easily pass an object through a solid. *Liquids:* You can see liquids and they do not have a specific shape. Liquids take on the shape of the container they are in. You can easily pass an object through a liquid. *Gases:* Gases are often invisible and can change their shape easily. Gases expand to fill whatever container they are in. You can easily pass an object through a gas.

Materials needed for each group:
3 resealable bags
colored water
a craft stick
a tennis ball

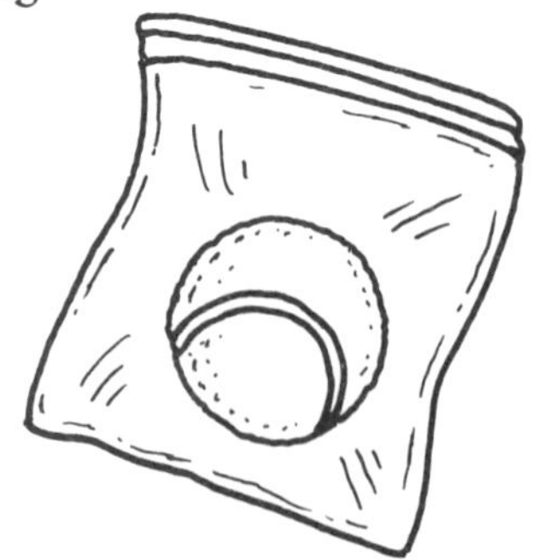

Directions for the activity:
Direct the students to place a tennis ball in one bag, put water in the second bag, and blow air in the third bag. Students will run three tests on each filled bag to determine whether it contains a solid, liquid, or gas.

1. Look at the contents of the bag.
2. Determine if the contents change shape easily.
3. Try to pass a craft stick through the contents of the bags.

As they work, the students complete the activity sheet "What's the Matter?"

Responding to Discovery

Discussion starters:

- Which bag contains a solid? How can you tell it is a solid?
- Which bag contains a liquid? How can you tell it is a liquid?
- Which bag contains a gas? How can you tell it is a gas?

Applications and Extensions

Ask students to name five different solids, five different liquids, and five different gases.

Real-World Applications

- Would you swim in a solid, a liquid, or a gas?
- Would you drink a solid, a liquid, or a gas?
- Would you sit on a solid, a liquid, or a gas?
- Is chewing gum a solid, a liquid, or a gas?
- Is the air you breathe a solid, a liquid, or a gas?

Name ______________________

What's the Matter?

Use your expert observation skills to learn about the different states of matter.

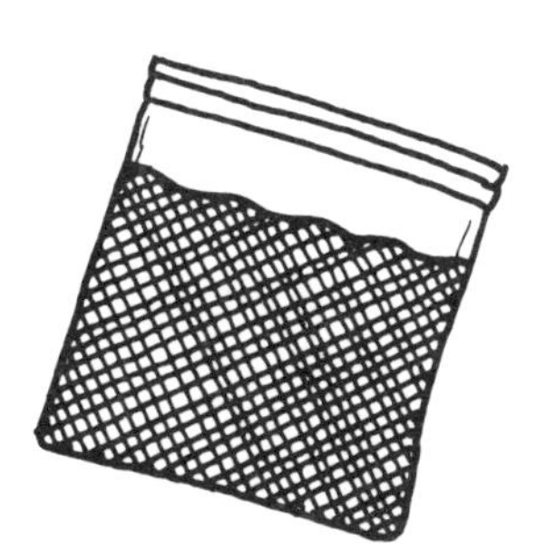

Bag 1: The water	Observations
Squeeze the bag. Does the water change shape?	
Open the bag. Can you pass a craft stick through the water?	
Can you see the water in the bag?	
What state of matter is the water?	

Bag 2: The tennis ball	Observations
Squeeze the bag. Does the tennis ball change shape?	
Can you pass a craft stick through the ball?	
Can you see the ball in the bag?	
What state of matter is the ball?	

Bag 3: The bag of air	Observations
Squeeze the bag. Does it change shape?	
Can you pass a craft stick through the bag of air?	
Can you see the air in the bag?	
What state of matter is the air?	

The Atoms Family

Gearing Up

Draw three boxes on the chalkboard. Label the boxes *Solid, Liquid,* and *Gas*. In the box labeled *Gas*, draw a few dots (atoms) scattered throughout. In the box labeled *Liquid*, draw several dots (atoms) distributed evenly. In the box labeled *Solid*, draw many dots tightly packed together. Have students study what you have drawn while recalling the previous discovery. Point out that the atoms in the solid are so close together that they cannot move. That is why a solid does not change shape and an object cannot pass through it. Discuss how the atoms in the solid, liquid, and gas differ.

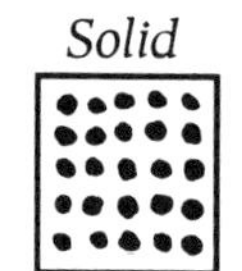

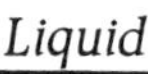

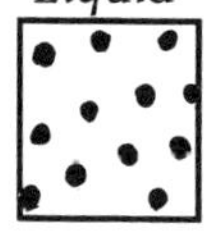

Gas

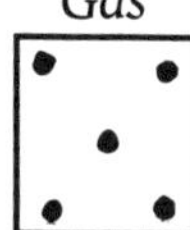

Process Skills Used

- making a model
- communicating
- classifying

Guided Discovery

Background information for the teacher:
Molecules are so small that we cannot see them. They are in constant motion whether they make up a solid, a liquid, or a gas. Atoms are the basic components of molecules. Atoms combine together to form molecules.

Directions for activity:
Explain to students that they are going to pretend that they are atoms. Atoms, just like children, are constantly moving. Everything is made up of atoms, but atoms in solids, liquids, and gases move differently.

Solids: Tell the students that the atoms in a solid are very close together. Have a group of students act like the atoms in a solid. They should group very closely in a small area and try to move together. Discuss how it felt to be an atom in a solid. How big were their movements inside the group?

Liquids: Explain that the atoms in a liquid are more spread out. Have students hold hands in groups of three as they move around a small section of the room. The space should be small enough that they touch the other groups as they move, but still move fairly freely. Discuss how it felt to be an atom in a liquid. How big were their movements inside the area?

Gases: Explain that a gas has no definite shape. Direct the students to move freely around the room. As they move, they may explore in every corner and against every wall, but they may not leave the room, unless you open the door or window. (You won't open the door because a gas fills whatever container it finds itself in.) Discuss how it felt to be an atom in a gas. How big were their movements inside the room?

Responding to Discovery

Divide the class into groups. Have each group of students demonstrate the behavior of the atoms in one state of matter. The rest of the class tries to guess what they are.

Applications and Extensions

Have students explain why it was easier to move as a liquid or a gas.

Real-World Applications

- Is all matter composed of atoms and molecules? How do you know?

Name ______________________________

The Atoms Family

Fill in each shape with plus signs (+) to illustrate the placement of the atoms.

- This key is a solid. Draw the atoms inside it.

- This glass of water is a liquid. Draw the atoms inside it.

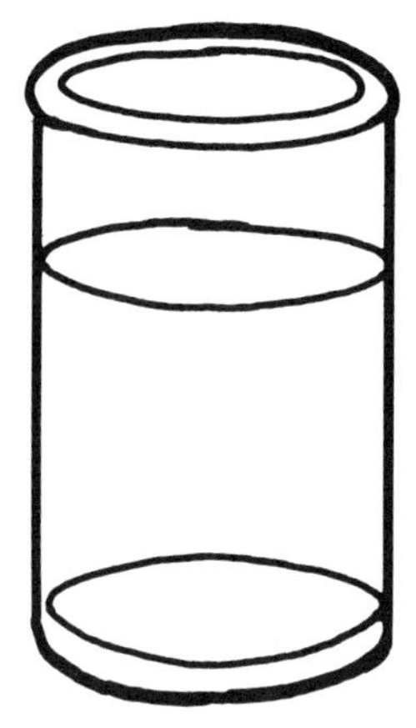

- This balloon is filled with gas. Draw the atoms inside it.

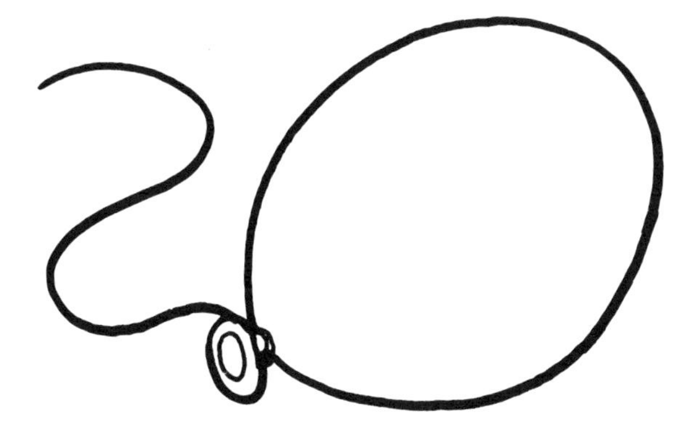

- Draw and label two other solids.

- Draw and label two other liquids.

- Draw and label two other objects filled with gas.

A Day at the Beach

Gearing Up

Hold a sheet of paper and ask the students what it is. Have them describe the attributes of the paper. (It is white, rectangular in shape, thin, a solid, etc.) Then, tear it in half quickly to grab the attention of the class. Ask the class to identify what it is now. Ask them to explain to you how it can still be paper after you changed it. Guide them to discover that the atoms that make up the paper are still atoms of paper. The state of matter has not changed, just the outside dimensions. Point out that this is called changing the matter physically.

Process Skills Used

- comparing
- observing
- predicting
- communicating

Guided Discovery

Background information for the teacher:
Matter can change physically or chemically. In this lesson, the students will observe a physical change. A physical change may alter the size or shape of the matter, but the type of matter stays the same. For example, a block of wood is still wood even if it is cut in half. You have only changed it physically. In a chemical change, a solid may become a liquid.

Materials needed:
a bag of sand (You can purchase a large bag from a toy or building supply store for only a few dollars.)
gallon-size (4 liter) resealable bags, different-shaped containers (e.g., milk cartons, coffee cans, frozen drink cans)
water
paper towels

Directions for the activity:
Give each group of students a gallon-size (4 liter) bag of sand and a container. Tell the students to dampen the sand with some water so it stays together. Allow the students to run their fingers through the sand to observe how it feels. Challenge the students to change the sand physically. Give the students time to mess around with different containers. They can pack the damp sand into a container and carefully turn it over onto a paper towel. As you monitor the work of the students, challenge them to return the sand to its original condition (put it back in the bag).

Responding to Discovery

As students complete the activity sheet, ask them what they discovered about the sand. When you changed the shape of the sand, was it still sand or did it become a different type of matter? What other types of matter can change physically? If you wrinkled a piece of paper, could you restore it to its original shape? What if you cut paper and glued it back together? Would that be an example of changing matter?

Applications and Extensions

Challenge the class to physically change other items. Blow up a balloon with a small amount of air. How can we physically change this balloon? (Add more air, let the air out, or pop the balloon.)

Real-World Applications

- List items made from wood.
- List items found or used at the beach.
- What states of matter are these items? How could they be changed physically?

Name ______________________

A Day at the Beach

- Predict how you can change the sand physically. ______________________

____

Draw the sand in the bag.

Draw the sand after you changed it.

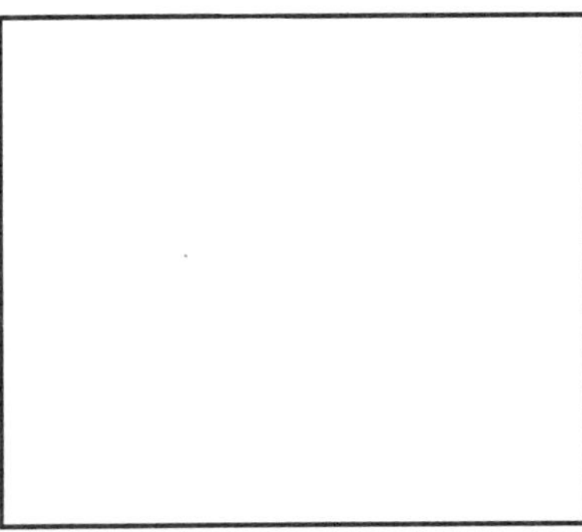

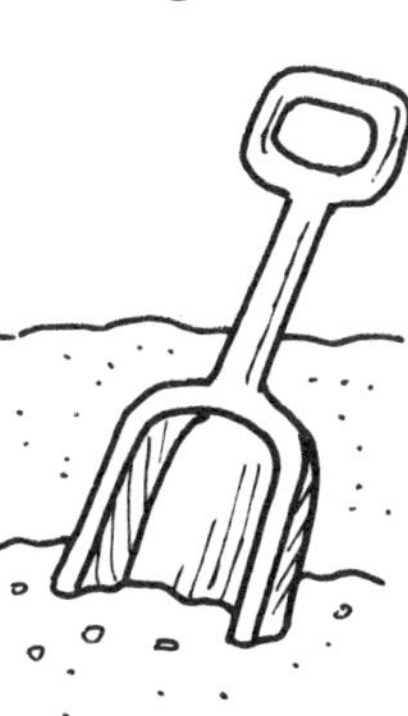

- I observed ____

____.

- Propose an explanation: A possible reason why this happened is ______

____.

- I have seen something like this happen before when ______________________

____.

Draw an item found or used at the beach in the box.
Determine whether the item is a solid, liquid, or gas.

Item ______________		Record your observation here
	Does the item change shape when you squeeze it?	
	Can you pass your finger through the item?	
	Can you see the item?	
	What state of matter is the item?	

Mold Me Again and Again

Gearing Up

Obtain a loaf of unsliced bread, a cutting board, and a serrated knife. Ask the students to determine the state of matter of the bread. Remind the students that a solid does not change shape easily. If you wish to change the shape of a solid, you often need a tool. Ask the students to propose how you can change the shape of the bread with this tool (the knife). Cut the bread into slices and share it with the class. When each student has a piece of bread, have them confirm that the slices are still bread although they do not look just like the original loaf.

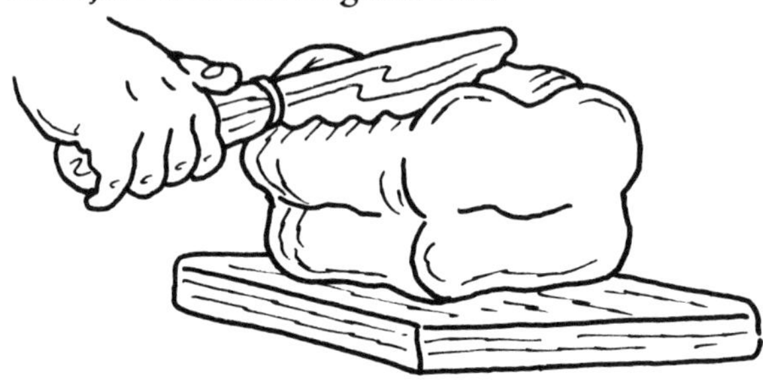

Process Skills Used

- observing
- communicating
- making a model

Guided Discovery

Background information for the teacher:
Through exploring with modeling clay, the students will build the understanding that matter can change physically and remain the same type of matter.

Materials needed:
a chunk of modeling clay for each student

Directions for the activity:
Before you distribute the clay, ask the students to describe its attributes. (What color is it? How does it feel? What shape is it? Describe its size.) Write their observations on the chalkboard. Have the students complete the activity sheet as they explore with their modeling clay. Guide the students as they roll their clay into a ball. Ask them to describe and write about its attributes. Then challenge them to return the clay to its original shape and size. Direct the students to explore how many ways they can change the clay physically.

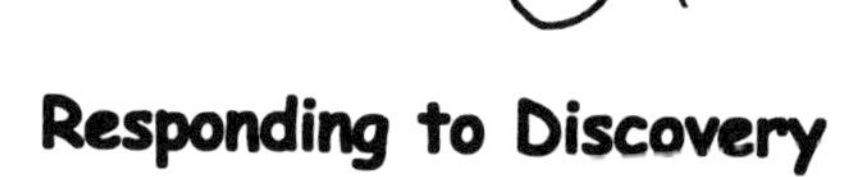

Responding to Discovery

- What are some shapes you made with the clay?
- At any time did the clay change from a solid to another state of matter?
- Could you always mold the clay back to its original shape?

Applications and Extensions

What other solid can you change physically? What tools do you need, if any?

Real-World Applications

- What physical changes have been (or can be) made to your house in the form of remodeling?

Name ______________________

Mold Me Again and Again

Roll the clay into a ball.
Draw your clay.

Describe its attributes.

____

____

Is it still a solid? ________ Is it still clay? ________

How do you know? ______________________

____

Flatten the clay.
Draw your clay.

Describe its attributes.

____

____

Is it still a solid? ________ Is it still clay? ________

How do you know? ______________________

____

Roll the clay into a snake.
Draw your clay.

Describe its attributes.

____

____

Is it still a solid? ________ Is it still clay? ________

How do you know? ______________________

____

Make your own shapes.
Draw your clay.

Describe its attributes.

____

____

____

____

A Watery Experience

Gearing Up

Ask the students, "What do you think would happen if you left a bowl of ice cream outside on a hot day?" (It would melt.) Ask them to hypothesize why the ice cream would melt. Guide the students to explain that the heat of the sun turns solid, frozen ice cream into liquid. Heat can change matter from one state to another.

Process Skills Used

- observing
- predicting
- comparing
- forming a hypothesis

Guided Discovery

Background information for the teacher:
Matter can change from one state to another because of various conditions. Heat can be applied to some solids to create a liquid. (This isn't always true. Heat applied to liquid cake mix turns it into a solid cake. Heat applied to paper does not create a liquid.) When heat is applied to ice, it changes from a solid to a liquid. Although we cannot see it, water vapor (a gas) is also released. In today's experiment, the ice cube on the black paper melts faster because black absorbs more heat than white.

Materials needed:
two ice cubes
white paper
black paper

Directions for the activity:
Display the materials and ask the students to predict what will happen when they place the ice cubes on the papers and set them in the sun. Acknowledge all responses. Then ask the students to predict which will melt faster—the ice cube on the black paper or on the white paper. They should record their predictions on the activity sheet, including the amount of time. You may take this activity outside or place the ice cubes in a sunny spot in the classroom. The students should record the time when they place the cubes on the papers and when the ice is completely melted on each paper. (Record a time for each paper separately.)

Responding to Discovery

Challenge the students to propose explanations for the outcome of the experiment. The students should feel the papers after the experiment and observe any differences in the papers. Challenge the students to propose other experiments they could try with the black and white paper in the sun. Carry out any experiments that are appropriate.

Applications and Extensions

Discuss what color of container might keep beverages cooler longer and why.

Real-World Applications

- Ask students where they have observed this before.
- Has anyone ever been too warm in the summer when wearing a black shirt?

Name ______________________

A Watery Experience

- Predict which ice cube will melt faster. ______________________
- Predict how long it will take. ______________________

Place the black and white papers in the sunshine. Place one ice cube on the black piece of paper and one on the white piece of paper. Watch the cubes and every five minutes write what you observed.

Ice cube on black paper		**Ice cube on white paper**	
Time	**Observation**	**Time**	**Observation**

- How many minutes did it take for the ice cube on the black paper to melt? ______________________
- How many minutes did it take for the ice cube on the white paper to melt? ______________________
- Explain the outcome of the experiment.

____

____

____

Blowing Bubbles

Gearing Up

Use a bubble wand to blow bubbles into your classroom. Ask the students to describe the bubbles. Tell the students that today we are going to make three different solutions of bubble stuff and compare the bubbles they produce.

Process Skills Used

- predicting
- classifying
- observing
- communicating
- measuring
- proposing explanations

Guided Discovery

Background information for the teacher:
A related concept to review in this exploration is that air has weight and takes up space.

Materials needed:
For the whole class: three different colors of dish soap at varying prices, corn syrup, and water
For each group: 3 paper or plastic cups, a plastic spoon, and a centimeter ruler
For each student: one straw and one plastic garbage bag

Directions for the activity:
Discuss the color, price, and appearance of each soap. Ask the students to predict which soap will produce the best bubbles.

Each group of students makes three bubble solutions. They should mix two spoonfuls of soap, two spoonfuls of water, and two spoonfuls of corn syrup in a cup. Repeat for each color of soap.

Cover each desk or table with a garbage bag. Demonstrate how to pour a puddle of soap on the plastic bag. To make a bubble, place the end of the straw in the puddle, lightly suck in to get soap into the straw, and then blow slowly and gently. When the bubble pops, measure the residue left behind in centimeters and record the results on the activity sheet. Repeat the procedure for each type of soap to determine which mixture blows the best bubbles.

Responding to Discovery

Discuss the results of the experiments. Ask the students to propose why one soap might create better bubbles than another.

Applications and Extensions

On commercials, dishwashing soap brands claim to cut grease. As an extension, observe how each soap reacts with oil. Add a drop of oil to a small puddle of water on the garbage bag. Notice how water and oil do not mix. Add a drop of each bubble mixture. How does the soap change the puddle? Does the soap that created the biggest bubbles mix best with the oil? Ask the students why their parents might choose to buy a soap that is more expensive. They may propose that the more expensive soap cleans better. Ironically, the least expensive soaps blow the best bubbles.

Real-World Applications

- Is soap still a liquid when it is in a bubble?
- Have you every tried cleaning a greasy pan with only water?
- Have you ever tried to get shortening off your hands? Why is this difficult?

Name ______________________________

Blowing Bubbles

- Predict which color soap will blow the best bubbles. Explain why you chose that color.

____

____

Color in each cup to match the color of each cup of soap.

Container of soap	Size of bubble in centimeters

- Which color soap made the biggest bubbles? ________________
- Was your prediction correct? ________________
- What shape are the bubbles? ________________
- What is inside a bubble? ________________
- Why are bubbles different sizes? ______________________________

____

- Is a bubble a liquid, a solid, or a gas? ________________ Explain.

____

____

____

Blow It Up

Gearing Up

Hold an empty balloon in front of the class. Ask the students to describe the balloon. They may tell you what it is made of, its color, what is in it, and its size and shape. Then ask them how to inflate the balloon. Follow their directions to inflate the balloon, then ask them to describe it and tell what is in it now. What gives the balloon its shape? Review the concept that air takes up space. In today's experiment, students will measure their lung capacity by filling balloons with air and measuring the circumference.

Process Skills Used

- observing
- measuring
- communicating

Guided Discovery

Background information for the teacher:
An expanded balloon is proof that air takes up space. The trapped air holds the balloon in an expanded state.

Materials needed:
one balloon per student and a centimeter tape for each group

Directions for the activity:
Before beginning the activity, remind students that for sanitary reasons they should put their mouths only on their own balloons.

First, allow the students to stretch their balloons by blowing them up and letting the air out. In each group, one student at a time takes a deep breath and blows all of the air in one breath into the balloon. The other members of the group can make sure that the student does not take another breath of air while blowing up the balloon. Then while the student holds the balloon closed, another student can measure the circumference (distance around the balloon). Students record the measurement in the group data table. Repeat with each member of the group for two measurements.

Responding to Discovery

Have students compare their lung capacity to that of their group members. Discuss ways of increasing lung capacity by increasing physical fitness. Also discuss what bad habits could produce having difficulty with this activity and why.

Applications and Extensions

Have students perform a physical activity such as running in place before checking lung capacity a third time. Determine whether physical exercise immediately affects lung capacity.

Real-World Applications

- Why are lungs important to us?
- What would happen if we did not have healthy lungs?
- What physical change occurred in the balloons?

Name ______________________________

Blow It Up

Names of team members	Circumference in centimeters		
	First trial	Second trial	Best attempt
1.			
2.			
3.			
4.			
5.			
6.			

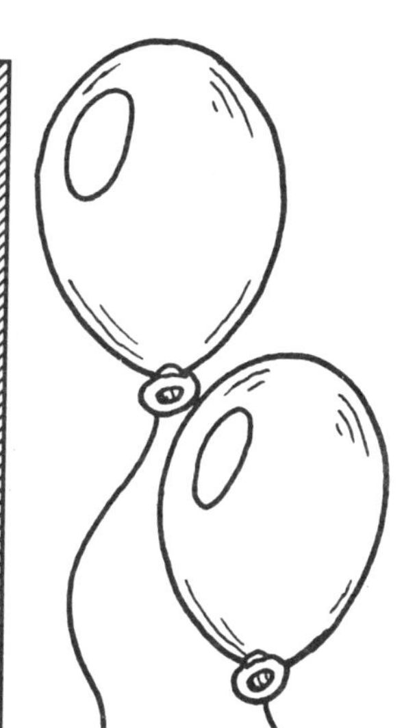

Plot the best attempts from the chart above onto the team graph below.

Circumference in centimeters						
24						
22						
20						
18						
16						
14						
12						
10						
8						
6						
4						
2						

Student names

- Which team member had the greatest lung capacity? ______________
- Write about the type of exercise you get each day.

____

____

Jet Propulsion

Gearing Up

Help the students create a mental picture of a jet airplane flying overhead. Ask them to hypothesize and discuss whether something is pushing the jet or pulling it. Tell the students that today we are going to explore jet propulsion. After the exploration, students will have more information to answer that question.

Process Skills Used
- predicting
- forming a hypothesis
- comparing
- communicating

Guided Discovery

Background information for the teacher:
An open balloon allows gas to escape. The balloon is propelled forward as the gas escapes out the back end. The amount of air in the balloon will affect how far the balloon will travel. This is similar to how rockets use jet propulsion to fly. The jet moves forward by reacting to the momentum of gaseous material being forced out the back of the engine.

Materials needed for each group of four students:
four different sizes of balloons
one plastic drinking straw
masking tape
a 25-foot-long (7½ m) string

Directions for the activity:
The teams must work cooperatively to complete the inquiry. Two people hold the string tightly at each end. On one end, thread the string through a straw. A third team member fully inflates the balloon and pinches the end so that the air does not escape. The fourth team member tapes the balloon to the straw using three pieces of tape. The opening of the balloon should face the person holding the string at that end. When the balloon is taped securely on the straw, the person holding the balloon releases it. Hold the string tightly and don't let go! Record observations.

Responding to Discovery

Students complete the activity sheet as they work. Encourage the students to design a variation of the inquiry to make the balloon move farther or at a faster rate. Their exploration should include a prediction and a record of events.

Applications and Extensions

Attach one end of the string to the ceiling and hold the balloon and straw at the other end of the string. Watch as the balloon propels upward.

Real-World Applications
- How does a jet move?
- What happens when you open a can of soda pop after shaking it? What causes this?

Name ____________________

Jet Propulsion

Problem: Does the amount of air in a balloon affect how far the balloon will travel?

Hypothesis: Make a hypothesis about which size balloon will travel farthest. ____________________

____

❧ Draw a picture of your inquiry and label each person and action in the picture.

Results: Record the distance of travel for each balloon.

____________________ ____________________

____________________ ____________________

Conclusion: Was your hypothesis supported? __________ Explain. ________

____

❧ Explain why that balloon traveled farthest. ____________________

____

❧ What pushed the balloon forward? ____________________

❧ When a jet is moving in the sky, is something pushing it or pulling it?

____

❧ How did you work as a team today? What might you do differently next time? ____________________

____

Making Ice Cream

Gearing Up

In a clear container, mix two colors or sizes of blocks or beads. Tell the students that you have mixed up your blocks and need help separating the mixture. Ask for suggestions on how to separate. In a glass, mix water and red food coloring. Ask the students for suggestions on how to separate the water and food coloring. It is impossible to separate this mixture because it has undergone a chemical change. Tell the students that today we will observe another chemical change when we mix liquid and solid ingredients to make ice cream.

Process Skills Used

- measuring
- observing
- communicating

Guided Discovery

Background information for the teacher:
Salt lowers the temperature at which the water freezes. As the ice melts, it draws heat away from the ice cream ingredients. The absence of heat causes the ingredients to freeze.

Materials needed for each group:
1 cup (240 ml) milk, 2 tsp. (10 ml) sugar, 1/2 tsp. (2½ ml) vanilla extract

1 gallon-size (4 liter) resealable bag and 1 quart-size (1 liter) resealable bag

1/2 cup (120 ml) table or kosher salt and 5 cups (1½ liters) of ice or snow

spoons and cups for serving ice cream

Directions for the activity:
In the small bag, place the milk, sugar, and vanilla. Teach the students to get rid of excess air and seal the bag. It is important that the bag is sealed. Shake the bag to mix the ingredients. Place the small bag into the large bag and layer ice and salt in the large

bag around the small bag. Then, seal the large bag. It is important that the bag is sealed. Team members take turns shaking the bags vigorously by holding the top of the bag with one hand and the bottom of the bag with the other hand. With continuous shaking, the ice cream will be hard in 6–8 minutes. When you open the bags, be careful not to get salt water in the small bag. Rinse the small bag before opening. Spoon into cups and enjoy.

Responding to Discovery

While making ice cream, discuss the individual ingredients. What other uses does each ingredient have? How did you change the milk physically? (Its shape changed when it was poured into the bag.) Once the ingredients are combined, the mixture changes chemically. When a chemical change takes place, a new product is formed—ice cream! Ask students what other chemical change took place. (The other chemical change occurs between the salt and ice.) Discuss how the salt caused the ice to melt. The ice drew heat from the ice cream and the students' hands—brr!

Applications and Extensions

The mixture of salt and water causes a chemical change, but over time, the two can be separated. Pour the salt water into a plastic plate and set it on the window sill for several days. When the water has evaporated—there is the salt!

Real-World Applications

- Think of a recipe that you can separate into its original ingredients after it is made.

Name ____________________

Making Ice Cream

Ice Cream
1 cup (240 ml) milk
2 tsp. (10 ml) sugar
½ tsp. (2½ ml) vanilla extract

- Pour the milk into the small bag. What physical change occurred?

____

____

- Mix in the other ingredients. Can you see the sugar anymore? How do you know it is still there? How could you test it?

____

Draw the small bag before shaking.

Now draw it after shaking.

- Predict how long it will take to make your ice cream the consistency of soft-serve ice cream. ____________________
- Put the small bag in the larger bag and add ice and salt. Shake the bags. Record how long it took to freeze. ____________________
- How could you return the solid ice cream to a liquid? ____________________

____

- How could you separate the sugar from the milk and vanilla? ____________________

____

- Describe a different recipe that involves a chemical change. ____________________

____

Let's Change the Temperature

Gearing Up

Purchase a chemical hand warmer from a sporting-goods store. Pass the cool hand warmer around the classroom before activating it and passing it around again. Have the students propose explanations for the spontaneous warmth. Discuss methods with which they are familiar for creating heat.

Process Skills Used

- comparing
- measuring
- observing
- communicating
- proposing explanations

Guided Discovery

Background information for the teacher:
A change in temperature can be a sign that a chemical change has taken place. Also, the calcium chloride (CaCl) cannot be removed from the water; therefore a chemical change has occurred.

Materials needed for each group:
2–6 Tbsp. (30 to 75 ml) calcium chloride (can be found in the supermarket near moth balls—brand name of Damp Rid)
plastic cup with 6 ounces (180 ml) of water
plastic spoon
thermometer

Directions for the activity:
Review safety rules regarding tasting or smelling chemicals. Students make observations about the temperature of the water in the cup by feeling the bottom of the cup and by measuring the temperature with a thermometer. Record the observations on the activity sheet.

Gently pour in 2 Tbsp. (30 ml) of the calcium chloride and stir while the thermometer is still in the water. After 3 minutes, check the temperature and record the results. Have each member feel the bottom of the cup to determine how the temperature has changed. Stir in 2 Tbsp. (30 ml) calcium chloride and stir again. Watch the thermometer. Repeat, if desired. **Caution: There might be a slight odor given off. Instruct students not to place their faces near the cup and to avoid smelling the odor.**

Responding to Discovery

Discuss the change that occurred. What happened when more calcium chloride was added? If heat is a form of energy, what must have been created in the chemical change (energy)? Do you think a gas was produced, too (yes)? Why do you think that (odor)?

Applications and Extensions

Observe how sodium chloride (rock salt) affects the temperature of water. (Enough salt will lower the temperature.)

Real-World Applications

- Chemicals are used in cold packs and heat packs.
- When might you use a hot or cold pack?

Name ______________________

Let's Change the Temperature

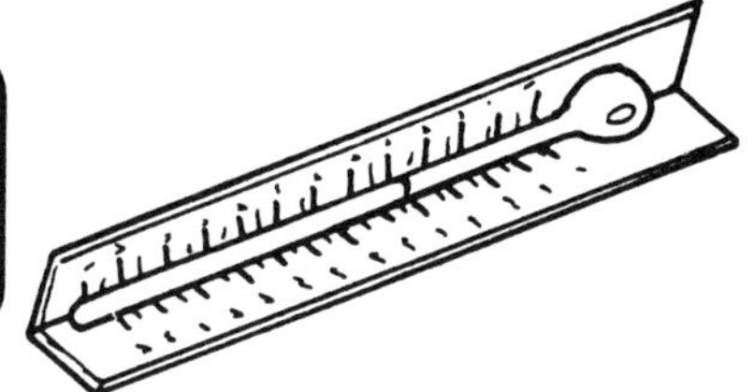

Problem: How will calcium chloride change the temperature of the water?

Hypothesis: Write a hypothesis about how calcium chloride will affect the water. ______________________

____

Record your observations about the temperature of each cup.

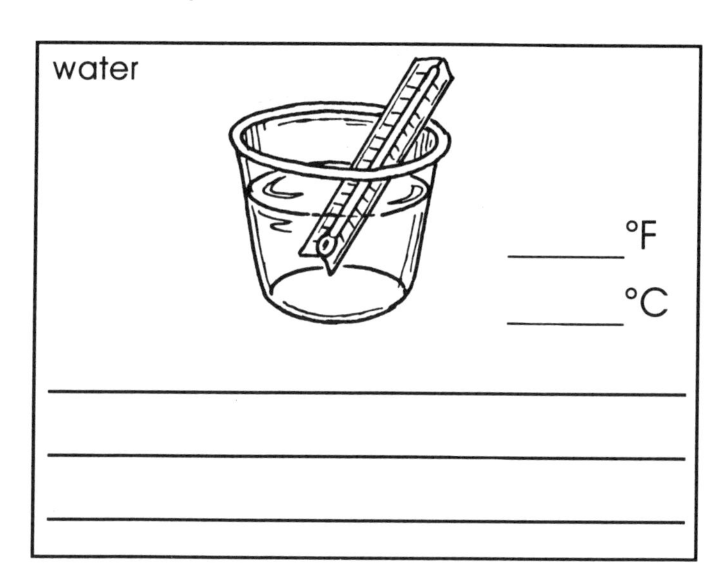

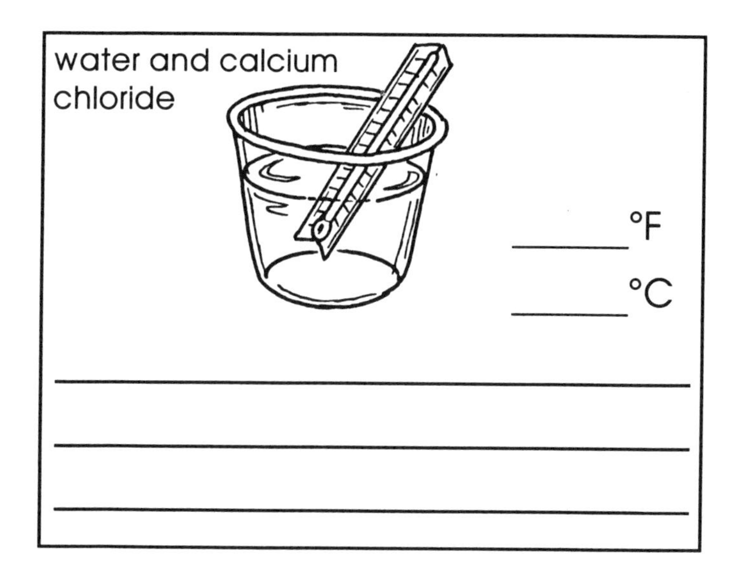

- Subtract the lower temperature from the higher one to show how much the temperature changed.

  ________ – ________ = ________

- **Results:** Did the liquid get colder or warmer? Write your answer in a complete sentence.

  ____

- **Conclusion:** What caused the change in temperature?

  ____

  ____

Cleaning Pennies

Gearing Up

We use cleaning products in our homes to keep them clean. These products are made of liquids that react with dirt to cause a chemical change. The chemical change removes the dirt. In this activity, you explore which liquid solution will best remove tarnish from pennies.

Process Skills Used

- observing
- predicting
- comparing
- controlling variables
- communicating

Guided Discovery

Background information for the teacher:
Pennies tarnish when oxygen chemically bonds to the surface. An acidic substance such as lemon juice or vinegar dissolves the bond. Add a little salt for even better results. For best results, use tarnished pennies from years prior to 1981. Pennies made before 1981 were made primarily of copper. Now pennies are made of copper-coated zinc.

Materials needed for each group:
small amounts of a variety of liquids (let the students propose the liquids) such as water, sugar water, vinegar, lemon juice, salt water, milk, and cool tea
5 tarnished pennies and paper towel

Directions for the activity:
Lids or caps from juice, milk, or other beverages as well as lids from jars work great for holding a small amount of liquid to clean pennies. Collect these ahead in a class recycling center. If they are unavailable, use paper cups or baby food jars. Before students begin exploring, they should predict which liquid will clean the pennies best. Pennies should be allowed to soak up to 10 minutes. Then, students can rinse and dry the pennies. They should record their observations on the activity sheet.

Responding to Discovery

Which solutions worked best for cleaning the pennies? Why do you think so? Did any solutions partially clean the pennies?

Applications and Extensions

To show the nature of chemicals in soda pops, have students place a piece of apple in different brands of soda to see which one will make it dissolve the fastest.

Real-World Applications

- Name three things you clean in your home.
- What type of cleaners do you use for each item? What do you think makes them work well?
- What are the active ingredients in different cleansers?

Name ______________________________

Cleaning Pennies

My Hypothesis:

- I predict that the ____________________ will best clean tarnished pennies.
- Choose five liquids. Place a penny in each. Draw each tarnished penny below and label the liquid in which it is soaking. Observe the pennies for up to 10 minutes.

☐ ☐ ☐ ☐ ☐

__________ __________ __________ __________ __________

- Draw and label each penny after the soaking time is up.

☐ ☐ ☐ ☐ ☐

__________ __________ __________ __________ __________

- **Procedure:** Write down the steps your group followed in this experiment.

First, ____

Next, ____

Then, ____

Finally, ____

- **Results:** Which solution worked best at cleaning tarnished pennies?

____

____

- **Conclusion:** Was your prediction correct? Do you think the solution that cleaned the pennies would be the best cleaner in all situations?

____

____

____

A Weighty Matter

Gearing Up

Ask a question such as "Which is heavier, a horse or a cow?" Listen to the students' answers and ask them to propose how to find the answer. Accept all reasonable answers, but listen for methods of weighing. Do the students have a concept of using a scale and of comparing mass? Tell the students that today we are going to explore finding the mass of solids.

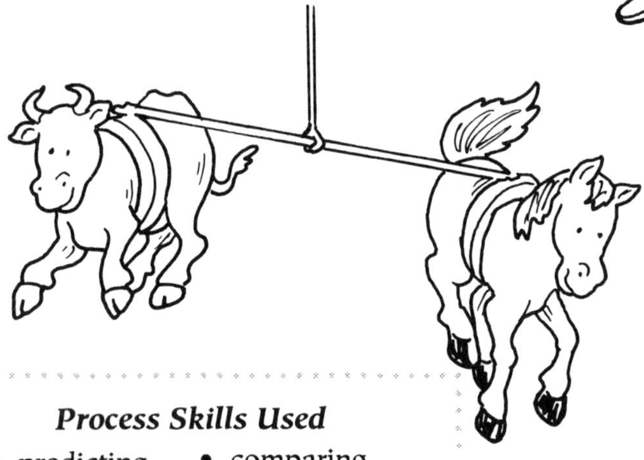

Process Skills Used

- predicting
- comparing
- measuring
- communicating

Guided Discovery

Background information for the teacher:
Mass is measured using units such as ounces, pounds, tons, grams, and kilograms.

Materials needed for each group:
a scale (preferably a balance scale)
six items to measure

scissors	pencil	battery
paper clip	calculator	marker
roll of tape	computer disc etc.	

1-gram, 5-gram, 10-gram, and 20-gram masses

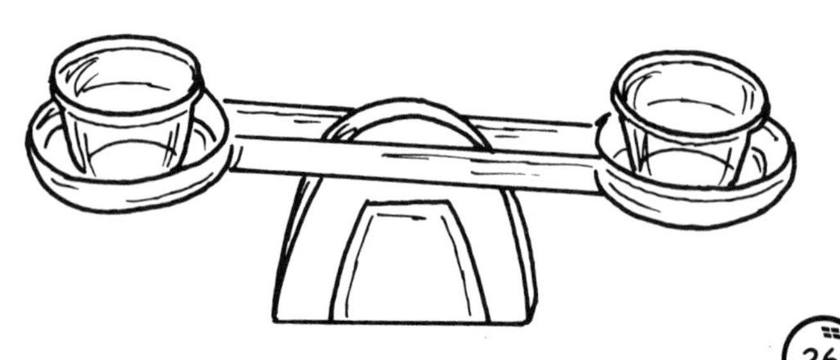

Directions for the activity:
Ask the students to order their objects from lightest to heaviest before measuring. Then have them measure and record the mass of each object. The students should complete the activity sheet as they work.

Responding to Discovery

Compare student predictions with actual measurements. Discuss when it may be necessary to know the actual mass of an object.

Applications and Extensions

Gather a variety of food and grocery items that have the mass printed on the packaging. Teach the students to calculate price per gram or ounce. Discuss value. Measure the mass of the items and determine the mass of the packaging.

Real-World Applications

- Students could obtain their height and weight records from their first years and compare.
- Discuss how doctors use this information in caring for them.

Name ______________________

A Weighty Matter

Predict: Draw the items in order from lightest to heaviest.

❧ As you measure the mass of each item, complete the table. Then, graph the results below.

Name of item	Weight in grams

Mass of objects

Mass in grams

26						
24						
22						
20						
18						
16						
14						
12						
10						
8						
6						
4						
2						

Items

❧ List the items in order from lightest to heaviest.

____

____

____

____

____

____

Liquid Layers

Gearing Up

Display a bottle of corn syrup and a bottle of rubbing alcohol. Ask the students to compare the liquids. Acknowledge their comparisons, then measure 60 mL, or ¼ cup, of each liquid into identical containers. Tell the students you want to find out which liquid is heavier. Ask for predictions. Place each container on a balance scale and measure in grams. Record the mass of each liquid. The students may be surprised that the liquids do not have the same mass. Pour the two liquids into a clear cylinder and observe. The two liquids will separate and the heavier liquid will sink to the bottom. Ask the students to propose explanations.

Process Skills Used

- classifying
- predicting
- measuring
- observing
- communicating
- proposing explanations

Guided Discovery

Background information for the teacher:
Density is the amount of matter packed into a given space. The corn syrup is more dense than the alcohol so the same volume of liquid has greater mass. In the following discovery, the student will compare the relative densities of various liquids. The liquids with greater density will sink below the liquids with lesser density.

Materials needed for each group:
cooking oil
water with food coloring
corn syrup
rubbing alcohol
a cylinder container such as an empty soda-pop bottle
¼ cup measuring cup or
60 mL measuring spoon

Directions for the activity:
Ask the students to predict and order their liquids from lightest to heaviest before measuring and mixing. Then have them measure equal amounts of each liquid and pour the liquids into the cylinder container. Students should complete the activity sheet as they work.

Responding to Discovery

Ask the students where they have seen something like this before.

Applications and Extensions

- Could gases be measured in a similar way?
- Drop some small solid objects into the layers of liquids. Observe in which layer each object comes to rest. Possible objects include a popcorn kernel, a piece of popped corn, a piece of cork or rubber, a slice of carrot, a marble, and a paper clip.

Real-World Applications

- Discuss the impact of oil spills and other pollutions in the oceans.

Name ______________________

Liquid Layers

- What do you think will happen if you pour the four liquids into the jar?

____

In what order did you add the liquids to the jar?

1.
2.
3.
4.

Draw the jar.

What was the final position of the liquids in the jar?

1.
2.
3.
4.

Draw the jar.

- Which liquid was the heaviest and sank to the bottom? ____________
- Which liquid was the lightest and floated on top? ____________
- A possible explanation for this result is ______________________

____.

- I have seen something like this happen before when ____________

____.

Does Air Have Weight?

Gearing Up

Obtain a helium-filled balloon. Release the balloon in the classroom for the students to observe. Ask the students to propose why the balloon floated to the ceiling. Helium appears to have no weight. Does air have weight? How can we find out? Discuss and listen to students' proposals.

Process Skills Used

- comparing
- observing
- making a model
- controlling variables
- communicating

Directions for the activity:
Demonstrate for the students how a ruler can act like a balance scale by balancing the ruler on two fingers. Tell the students that they will tape a balloon on each end of the ruler and compare the mass of each balloon. One balloon will be empty and the other will be filled with air. Students should complete the activity sheet as they work.

Guided Discovery

Background information for the teacher:
The helium in the balloon is less dense than the air surrounding the balloon so the helium-filled balloon rises. The density of the air inside an air-filled balloon is the same as the air outside the balloon. The latex balloon itself is heavy enough to drag the balloon down to the ground. In today's discovery, the students will observe that air does have weight when they compare an empty balloon with an air-filled balloon.

Materials needed for each group:
two identical balloons
tape
a ruler or meter stick

Responding to Discovery

Drop two different-sized inflated balloons to see if the one with more air will fall faster than the one with less air. Discuss whether this experiment will prove anything.

Applications and Extensions

Have the students propose a discovery that will prove that air has weight.

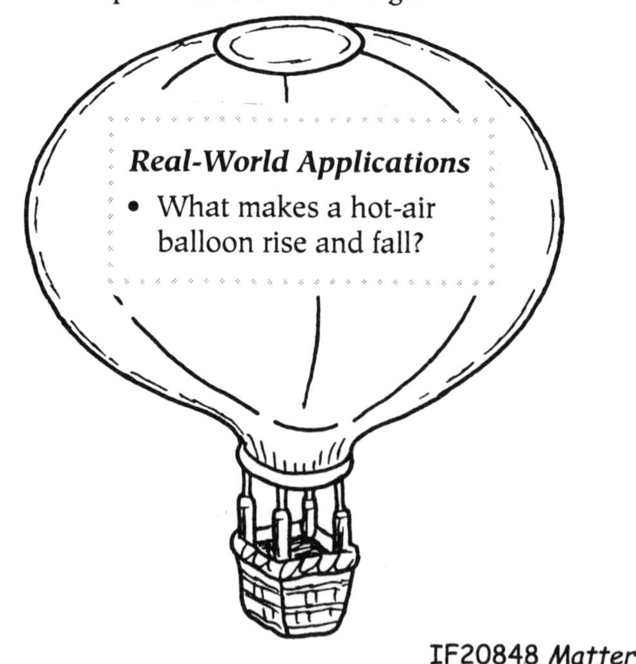

Real-World Applications

- What makes a hot-air balloon rise and fall?

Name ___________________________

Does Air Have Weight?

- Compare the two balloons. They should be identical. Blow up one balloon and tie it closed. Draw both balloons.

- Predict which balloon weighs more. ___________________________

- Draw a picture of the balloons balanced on the ruler. (Draw yourself in the picture.) Label all the parts.

- Propose an explanation. Write a possible reason why what you observed happened.

____

____

- What are some ways that we use air? ___________________________

____

____

Performance-Based Assessment

3 = Exceeds expectations
2 = Consistently meets expectations
1 = Below expectations

Student Names													
Lesson Investigation Discovery													
Lesson 1: What's the Matter?													
Lesson 2: The Atoms Family													
Lesson 3: A Day at the Beach													
Lesson 4: Mold Me Again and Again													
Lesson 5: A Watery Experience													
Lesson 6: Blowing Bubbles													
Lesson 7: Blow It Up													
Lesson 8: Jet Propulsion													
Lesson 9: Making Ice Cream													
Lesson 10: Let's Change the Temperature													
Lesson 11: Cleaning Pennies													
Lesson 12: A Weighty Matter													
Lesson 13: Liquid Layers													
Lesson 14: Does Air Have Weight?													
Specific Lesson Skills													
Can make reasonable predictions.													
Can make detailed observations.													
Can propose an explanation.													
Can follow written directions.													
Can use a ruler and measure to the nearest centimeter.													
Can work cooperatively with a partner or group.													
Can build on observations by asking appropriate questions.													
Can create a graph based on data from investigations.													
Can use a magnifying glass.													
Can read a thermometer.													
Can find mass to nearest gram.													
Can communicate through writing, drawing, and dialogue.													
Can apply what is learned to real-world situations.													